Laís Monique Gomes do Ó
Alfredo M. De Sousa
Tiago C. Da Silva

Morphological characteristics of cowpeas

Laís Monique Gomes do Ó
Alfredo M. De Sousa
Tiago C. Da Silva

Morphological characteristics of cowpeas

Subjected to salinity levels

ScienciaScripts

Imprint

Cover image: www.ingimage.com

This book is a translation from the original published under ISBN 978-3-330-76704-1.

Publisher:
Sciencia Scripts
is a trademark of
Dodo Books Indian Ocean Ltd. and OmniScriptum S.R.L publishing group

120 High Road, East Finchley, London, N2 9ED, United Kingdom
Str. Armeneasca 28/1, office 1, Chisinau MD-2012, Republic of Moldova, Europe
Managing Directors: Ieva Konstantinova, Victoria Ursu
info@omniscriptum.com

Printed at: see last page
ISBN: 978-620-8-55137-7

PREFACE

The cowpea crop is of great socio-economic importance to the north-eastern region of Brazil, especially in the context of family farming.

Over the years, various studies have been carried out investigating the effect of different factors that influence the crop's production system. Among the factors investigated, research related to the development of this crop when subjected to saline conditions in irrigation water has gained a lot of prominence.

In this respect, this book aims to become a source of reference for students and researchers, containing references to important work already carried out in the area and presenting the results of a study carried out with this crop under levels of irrigation water salinity.

We wish you a good reading and hope that in some way we are contributing to scientific practice in this linc of agronomic research.

The Authors

ACKNOWLEDGEMENTS

To GOD Almighty, for everything, especially for the faith that everything would happen according to his will.

To my parents and brother for their support and encouragement at all times.

To my family, especially my aunt Sheila, my cousin Mayara and my beloved grandmother Pretinha for always believing in my efforts and for their financial help.

To my husband Rafael José for his patience and love for me, without all of this it would have been a difficult journey.

To the Federal University of Ceará (UFC), for believing in the professional growth that comes from knowledge and for the opportunity granted.

To CNPq for awarding me the scholarship that greatly contributed to the completion of the course.

To my first supervisor, Prof Dr José Carlos de Araújo, for the opportunity he gave me, which opened up new knowledge.

To Professor Dr Thales Vinicius de Araújo Viana for his friendship, for his guidance during the two years I was his scholarship student and for the opportunity to be a tutor in his course.

To Prof Dr Francisco Marcus Lima Bezerra, for welcoming me with open arms when I needed it most, for his excellent guidance, trust and friendship.

To my co-supervisor João Valdenor Ferreira Filho for his

contributions, his friendship and his efforts to ensure that this work turned out as expected.

To the member of the examining board Francisca Robevania Medeiros Borges for her time, valuable contributions, reflections, suggestions, support, true friendship and many moments of laughter.

To my dear friend and member of the examining board, Keivia Lino Chagas, for the hours we spent studying together, for her friendship, her words of encouragement and all her support during these unforgettable times as an undergraduate.

To my undergraduate friends Raphael Jatahy, Fabricio Martins, Tiago Cavalcanti, Carol Bayma, Marcela Caroline, Wladislaw Cavalcanti, Wladmir Stênio, Gabriela Barroso, Valsergio Barros, José Israel and the rest of my classmates for their friendship and for sharing the information that helped me get this far.

To my research friends Amanda, Chrislene, André Rufino, Newdmar, Emília, Ana Paula, Alexandre, Geocleber, Mário, Bruno, Luiz Gonzaga and Roney for their encouragement.

To the staff and professors of the Department of Agricultural Engineering, who contributed in some way to my education.

To everyone who contributed directly or indirectly to the completion of this work.

Devote everything you do to the Lord and *your plans will* ***succeed"*** *Proverbs 16:3*

SUMMARY

The use of irrigation is important for the viability of commercial exploitation of the bean crop, especially in the semi-arid region of the Northeast, due to the irregularity of rainfall. However, inadequate irrigation management has led to soil salinisation, causing reductions in the morphological characteristics of crops in general. With this in mind, the aim was to assess the effects of different levels of irrigation water salinity on the morphological behaviour of two cowpea cultivars. The experiment was conducted in open-air pots in the experimental area of the Agrometereological Station of the Federal University of Ceará, Fortaleza, Ceará. A split-plot randomised block design was used, with the salinity levels distributed in the plots and the cultivars in the sub-plots. The salinity levels corresponded to: S1, salinity level of 0.7 dS m^{-1} (water from a groundwater well belonging to the Agrometeorological Station of the Federal University of Ceará); S2, salinity level of 1.5 dS m^{-1}; S3, salinity level of 3.0 dS m^{-1}; S4, salinity level of 4.5 dS m^{-1}; S5, salinity level of 6.0 dS $m^{(-1)}$. The cultivars evaluated were Epace 10 and BRS Itaim. The seeds were planted in 15 litre pots containing sieved soil and organic fertiliser. The variables assessed were: plant height, number of leaves, stem diameter and leaf area. The results were analysed by variance and the means were compared using the Tukey test. The plant height, number of leaves and leaf area of the bean plant were reduced with increasing levels of salinity in the irrigation water. Among the cultivars evaluated, the Epace 10 cultivar showed higher values for plant height and leaf area.

Keywords: Vigna unguiculata (L) Walp. Morphological characteristics. Cultivars. Water quality.

SUMMARY

CHAPTER 1

INTRODUCTION

Cowpea (Vigna unguiculata L.) is the main subsistence crop and source of plant-based protein for small producers in the semi-arid region of the northeast of Brazil. Because of its hardiness and precociousness, it is considered a plant adapted to the critical conditions of the semi-arid region, and is widely grown in rainfed conditions and irrigated perimeters.

The use of irrigation in north-eastern Brazil is important for the viability of commercial exploitation of the bean crop, due to the irregularity of rainfall. On the other hand, irrigation has caused some problems for the environment. One example of this is the use of inferior quality water for irrigation, most of which is saline, causing a drop in crop yields. This makes it necessary to adopt appropriate crop, soil and irrigation water management practices.

Among the management practices recommended to produce satisfactorily in soil and water conditions with a high risk of salinisation are the use of plants tolerant to salinity and sodicity, proper irrigation management and soil salinity control through the application of leaching slates, making it important to carry out studies aimed at assessing the sensitivity of species to saline stress (TRAVASSOS et al., 2012).

In this context, the main objective of this work is to evaluate the growth of cowpea plants subjected to salinity levels in irrigation water with the BRS Itaim and Epace 10 cultivars.

CHAPTER 2

LITERATURE REVIEW

2.1 General aspects

The cowpea's probable centre of origin is Africa, but it grows in a wide environmental range, from latitude 40° N to 30° S, adapting to both high and lowlands in West Africa, Asia, Latin America and North America (CARDOSO et al., 2005; DADSON et al., 2005). It was domesticated in agricultural systems composed of sorghum and millet, which predominate in the semi-arid regions of West Africa, which is characterised by having one of the largest areas of cowpea production (SINGH et al., 2002).

The cowpea goes by various common names and is known as "feijão-de-corda", "feijão-macassar" or "feijão-caupi" in the Northeast; "feijão-da-colônia" and "feijão-de-praia" in the North and "feijão-miúdo" in the South. In some regions of Bahia and the north of Minas Gerais, it is known as catador bean and gurutuba bean and as fradinho bean in the states of Bahia and Rio de Janeiro (FREIRE FILHO et. al., 2000).

Although the use of the cowpea is very similar to that of the common bean (Phaseolus vulgaris L.), it is considered to be a legume with a wide distribution worldwide, being found mainly in the semi-arid, humid and sub-humid tropics, climatic conditions to which it is best adapted (SMARTT, 1990; MOUSINHO, 2005).

The cowpea is an annual herbaceous plant that is propagated by seeds. It is self-fecundating and has a very low rate of allogamy or cross-fertilisation. The flowers are complete, with the male and female organs well protected by the five petals, which are white, yellow or violet in colour. They have ten stamens, the stipe ends in a recurved stigma that is moist and covered in hairs in order to better adhere the pollen grains. The ovary is narrow and elongated and anthesis normally takes place in the early hours of the morning, which means that pollination occurs predominantly through insects (TEÓFILO; MAMEDE; SOMBRA, 1999). The root system is pivotal, reaching up to 0.80 metres deep, and like other legumes, it has the ability to fix nitrogen from the air (MOUSINHO, 2005).

It is considered a plant with a C3 photosynthetic metabolism and has two types of growth habit, determinate and indeterminate, the latter being characteristic of the most predominant genotype cultivated in Brazil, which differs from determinate in that the stem does not produce a limited number of nodes, continuing to grow even after the inflorescence has been issued (ARAÚJO et al., 1981). According to Freire Filho et al. (2000), cowpeas can be classified in terms of cycle as follows: super-early cycle (maturity is reached up to 60 days after sowing); early cycle (maturity is reached between 61 *and* 70 days after sowing); medium cycle (maturity is reached between 71 and 90 days after sowing); late cycle (maturity is reached from 91 days after sowing).

2.2 Salt stress

The high concentration of salts is a stress factor for plants, as it has an osmotic activity that retains water, in addition to the action of ions on the protoplasm. Ayers and Westcot (1999) state that the ions that most often cause toxicity are chloride, sodium and boron. An excess of Na^+ and, above all, an excess of Cl^- in the protoplasm cause disturbances in the ionic balance (K^+ and Ca^{2+} in relation to Na^+), as well as the specific effect of ions on enzymes and membranes (LARCHER, 2006).

In the north-east of Brazil, most of the water used for irrigation has a salt concentration in the range of 1 to 30 mmolc L^{-1}, corresponding to the electrical conductivity range of 0.1 to 3.0 dS m^{-1}, and its use in agriculture, combined with intense evaporation and the lack of adequate soil drainage, can cause salinity problems (HOLANDA; AMORIM, 1997). As a result, checking the suitability of water for irrigation purposes should mainly consider factors related to the content and type of salts, soil characteristics, crop tolerance, climatic conditions, irrigation management and drainage conditions (RHOADES et al., 2000).

Salinity in irrigated areas occurs as a result of the use of water of inadequate quality, associated with the management of the soil-water- plant system and, whatever its source, the water used for irrigation always contains salts, although the quantity and quality of the salts present vary greatly (MEDEIROS; GHEYI, 1994).

The effect of salinity on plants appears primarily in the diversion of energy from the growth process to maintain osmotic potential. One of the first processes from which energy is diverted is cell elongation. The cells in leaf tissues divide but do not elongate. The occurrence of more cells per unit area causes them to appear dark green in colour, a typical symptom of an osmotically stressed plant. In addition, the osmotic effect will depend on the plant's sensitivity to a specific ion in the irrigation water or solution, boron toxicity being the most common (BOHN et al., 1979).

According to Tyerman and Skerrett (1999), in saline environments, sodium chloride (NaCl) has been shown to be the predominant salt, causing most of the damage to plants. The osmotic effect has been the main cause of reduced plant growth, associated with ion toxicity due to the excessive absorption of sodium and chlorine and the nutritional imbalance caused by disturbances in the absorption and/or distribution of essential nutrients (YAHYA, 1998).

Sodium can act by stimulating plant development, as it is considered a useful element for many crops due to its own physiological effects or because it replaces part of the K^+ required by the plant. The degree of this substitution will depend on the absorption potential and translocation of Na^+ to the aerial part of the plant, as well as the plant's efficiency in utilising K^+ (SILVA, 1998).

2.3 General effects of salt stress on bean plants

Agriculture is facing a major problem worldwide with the lack of adequate water resources, forcing many producers to use brackish water for crop irrigation (REED, 1996). In many production areas, the use of low quality water for irrigation and the application of excessive amounts of fertiliser are the main causes of soil salinisation. In semi-arid regions this is a serious problem, as it results in limited agricultural production and reduced productivity of important crops.

Salt stress inhibits plant growth by reducing the osmotic potential of the soil solution, restricting the availability of water, as well as causing excessive accumulation of ions in plant tissues, which can cause ion toxicity, nutritional imbalance or both. However, the degree of severity with which each of these components influences plant growth and development depends on many factors, including: the plant species or cultivar, the phenological stage, the saline composition of the medium, the intensity and duration of the stress, as well as soil and climate conditions and irrigation management (YEO, 1999; SILVA et al., 2003; GHEYI et al., 2005).

Plant growth processes are particularly sensitive to the effect of salts, so growth rate and biomass production are good criteria for assessing the degree of stress and the plant's ability to overcome salt stress (MORAIS et al., 2011).

In order to withstand salt stress, plants have developed complex mechanisms that help them adapt to the osmotic and ionic stresses caused by soil salinisation. It should be noted that genotypes with different degrees of tolerance to salt stress probably reflect the variability in the efficiency of acclimatisation or adaptation mechanisms to excess salts in the soil (COSTA et al. 2003; DIAS; BLANCO, 2010).

One of the mechanisms by which plants adapt to salt stress is a reduction in leaf area or the loss of leaves through abscission. Leaves are the most sensitive plant structures when it comes to perceiving environmental stresses. Those that are capable of osmotic adjustment maintain turgidity at lower water potentials, which allows them to continue growing and facilitates the maintenance of stomatal opening for longer. Inorganic ions are stored in the vacuole so as not to interfere with the metabolism of the cytoplasm, while organic solutes are compartmentalised in the cytosol so as to maintain the water balance between the vacuole and the cytoplasm (TAIZ; ZEIGER, 2009).

Another mechanism used by plants in conditions of high salinity is osmotic adjustment (TAIZ; ZEIGER, 2009). As osmotic adjustment is a process by which the leaf's water potential can be lowered without a decrease in turgor or cell volume, it is therefore a net increase in the content of solutes per cell. Most osmotic adjustment can be calculated by increasing the concentration of a variety of common solutes, including sugars, organic acids, amino acids and inorganic ions, especially K^+ (TAIZ; ZEIGER, 2009).

Some crops, due to their better osmotic adaptation capacity, are able to produce satisfactorily even in saline environments. This adaptability is very useful and allows for the selection of the most tolerant crops, capable of producing economically acceptable yields, when soil salinity cannot be kept below the tolerance level of traditionally cultivated plants. The maximum level of average salinity in the root zone that plants can tolerate without any negative effect on their development and yield is known as the threshold salinity (SL), which is usually expressed in terms of ECes

(GHEYI et al., 1991; AYERS and WESTCOT, 1999).

According to Maas and Hoffman (1977), cowpea is a species considered to be moderately sensitive to salt stress. However, the literature has shown that this species tolerates saline water with an electrical conductivity of up to 3.3 dS m^{-1} and threshold soil salinity (ECes) of around 4.9 dS m^{-1}, making it moderately tolerant (AYERS; WESTCOT, 1999). The concentrations of salts that restrict the growth of string beans vary between cultivars and seem to depend on the ionic composition of the medium (LACERDA, 1995).

CHAPTER 3

MATERIAL AND METHODS

3.1 Location and characterisation of the experimental area

The experiment was conducted in the experimental area of the Agrometeorological Station of the Agricultural Engineering Department of the Agricultural Sciences Centre of the Federal University of Ceará (UFC), in the municipality of Fortaleza, state of Ceará, geographical coordinates 03°45'S, 38°33'W, average altitude 19 m, from October to December 2014 (Figure 1).

Figure 1 - Location of the experimental area. Agrometeorological Station, Fortaleza, Ceará, 2014

Source: Authors

According to the Koppen classification, the region's climate is Aw'>

tropical rainy, with high temperatures and a rainy season predominantly in autumn. The average monthly values of the climatic variables during the experiment are shown in Table 1.

Table 1- Average weather conditions observed during the experiment. Agrometeorological Station, Fortaleza, Ceará, 2014

Month	Temperature Air Average	Relative humidity	Speed Wind	Precipitation
	(°C)	(%)	($m\ s^{-1}$)	(mm)
October	27,4	68	4,5	6,6
November	27,2	71	4,3	17,7
December	27,5	70	4,1	53,5

The substrate used to grow the plants was a sample of Argissolo Vermelho-Amarelo (EMBRAPA, 2013), from an area with native vegetation, located near the experimental area of the Agrometeorological Station of the Department of Agricultural Engineering, in the municipality of Fortaleza, collected at a depth of 0 - 0.20 metres.

When the pots were filled, a composite soil sample was taken to characterise the soil's physical and chemical attributes and sent to the soil and water laboratory at the Federal University of Ceará, the results of which are shown in Tables 2 and 3, respectively.

Table 2 - Physical and water attributes of the soil used as a substrate for growing the plants in the experiment. Agrometeorological Station, Fortaleza, Ceará, 2014

Layer	Particle size composition			Textural class	Density		Humidity	
	Sand	Silt	Clay		Soil	Particles	33 Kpa	1,500 Kpa
(m)	(g kg-1)			(-)	(g cm-3)		(g /100g)	
0,00 - 0,20	785	76	139	Sandy loam	1,49	2,55	7,51	4,77

Table 3 - Chemical attributes of the soil used as a substrate for growing the plants in the experiment. Agrometeorological Station, Fortaleza, Ceará, 2014

Layer	pH (Water)	EC	Assortative Complex					
			Ca^{2+}	$Mg2^{+}$	K^{+}	In^{+}	$H^{+} + Al^{3+}$	Al^{3+}
(m)	(-)	(dS m'1)	(cmolc kg'1)					
0,00 - 0,20	4,2	0,17	0,70	0,60	0,15	0,06	3,14	0,95
Layer	S	T	V	PST	M.O	P Assimilable		
(m)	(cmolc kg'1)		(%)		(g kg-1)	(mg kg'1)		
0,00 - 0,20	1,5	4,6	33	1,0	9,41	3		

pH - hydrogen potential, EC - electrical conductivity of the soil saturation extract; S - sum of bases; T - cation exchange capacity; V - base saturation; PST - percentage of exchangeable sodium; O.M. - organic matter.

3.2 Plant material

Two bean cultivars were used (Epace 10 and BRS Itaim), produced by the Agricultural Research Company of Ceará and Embrapa Meio Norte.

3.2.1 Epace 10

Launched in 1988 by the Agricultural Research Company of Ceará

(EPACE), the Epace 10 cultivar was the result of the CNCx166-8E line obtained from the Seridó x TVu 1888 cross carried out at CNPAF/EMBRAPA. It has a cycle of 65 to 75 days, semi-branching, violet flowers and brown seeds, with an average yield of 1,000 kg ha^{-1}. Figure 2A shows the aerial part of this cultivar's plant and figure 2B shows the cowpea grains.

Figura 2 - Photos of the Epace 10 cultivar. (A) Aerial part of the Epace 10 cultivar grown in pots and (B) Cowpea grains. Agrometeorological Station, Fortaleza, Ceará, 2014

3.2.2 BRS Itaim

The BRS Itaim cultivar corresponds to the MNC04-786B-87-2 strain, obtained from the MNC04-786 cross made at Embrapa Meio Norte in 2004. It has a cycle of 60 to 65 days, erect growth, white flowers with slight purple pigmentation, white seeds and a black hilum, with an average grain yield of 1,618 kg ha^{-1}. Figure 3A shows the aerial part of the plant (cultivar BRS Itaim) and figure 2B shows the grains of this same cultivar of cowpea.

Figura 3 - Photos of the BRS Itaim cultivar: (A) Aerial part grown in pots and (B) cowpea grains. Agrometeorological Station, Fortaleza, Ceará, 2014

3.3 Description of the experimental area

The total area occupied by the cowpeas was 12.0 x 3.0 m = 36.0 m^2, where the 90 pots (plastic buckets) were installed, with a unit volume of 15 litres, distributed in the space according to Figure 4, with 0.8 m between rows and 0.5 m between pots (plants).

Figura 4 - General view of the experimental area with details of the plots (A) and sub-plots (B). Agrometeorological Station, Fortaleza, Ceará, 2014

3.4 Setting up and conducting the experiment

Initially, before depositing the soil used as a substrate for growing the plants, a layer of gravel No. 1 was placed at the bottom of the pots to facilitate free drainage of excess water, which was then filled in and covered with a blanket to prevent possible soil loss. The flexible plastic pots, with a volume capacity of 15 litres, had holes at the bottom to remove any excess water.

The pots were then filled with a substrate made up of a mixture of local soil and organic compost from worm humus in a 2:1 ratio (Figure 5).

Figura 5 - Detail of the distribution of filled pots in the experimental area. Agrometeorological Station, Fortaleza, Ceará, 2014

3.5 Fertilising, sowing and thinning

The chemical fertilisers for the foundation were applied based on the soil analysis and the nutritional requirements of the crop, following the prescriptions of Aquino et al. (1993). The quantities of macronutrients required were as follows: 20 kg ha^{-1} of nitrogen (N), 80 kg ha^{-1} of phosphorus (P) and 30 kg ha^{-1} of potassium (K), using urea, monoammonium phosphate (MAP) and potassium chloride, respectively. However, at the time of planting, only the amounts of MAP and potassium chloride were applied in their entirety, equivalent to 15 and 5 g pot^{-1}. The urea source was applied in equal instalments and at 15 and 30 DAP as a top dressing, the amount corresponding to 5 g pot^{-1}.

In order to correct and/or avoid possible micronutrient deficiencies, the equivalent of 10g of the commercial formulation FTE BR 12 [Calcium (Ca): 7.1%; Sulphur (S): 5.7%; Boron (B): 1.8%; Copper (Cu): 0.8%; Manganese (Mn): 2.0%; Molybdenum (Mo): 0.1% and Zinc (Zn): 9.0%] was also applied at planting time.

Planting was done by sowing an average of three to four seeds per pot. At 7 days after sowing, germination was approximately 90 per cent. In the thinning, carried out manually at 10 DAS, the plants were pulled out close to the ground, leaving two plants per pot, in order to obtain the plant stand desired by the experiment (Figure 6).

Figura 6 - Detail of the two plants left per pot when thinning. Agrometeorological Station, Fortaleza, Ceará, 2014

3.6 Experimental design and composition of treatments

The statistical design was randomised blocks with split plots. The salinity levels were distributed in the plots and the cultivars in the subplots.

The salinity levels defined were: S1, salinity level of 0.7 dS m^{-1} (water from a groundwater well belonging to the Agrometeorological Station of the Federal University of Ceará); S2, salinity level of 1.5 dS m^{-1}; S3, salinity level of 3.0 dS m^{-1}; S4, salinity level of 4.5 dS m^{-1}; S5, salinity level of 6.0 dS $m^{(-1)}$.

Each plot was divided into two subplots in which the cultivars were randomly distributed. According to the sub-plots drawn, one of the cultivars chosen was distributed in the first three pots and the other three pots chosen were distributed in the other three pots. Figure 6 shows a sketch of the experimental area.

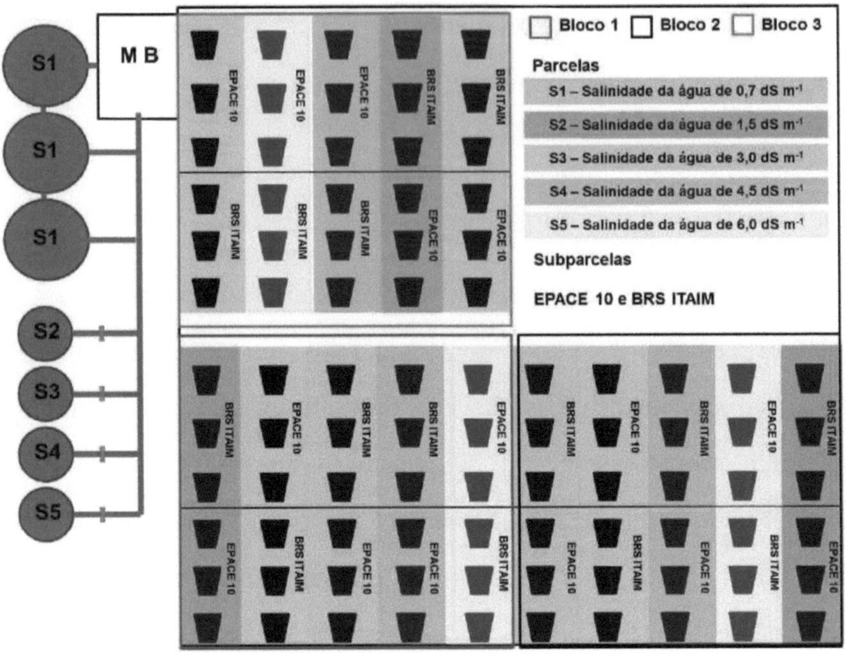

Figura 7 - Sketch of the experimental area, showing the layout of the experimental units. Agrometeorological Centre, Fortaleza, Ceará, 2014.

3.7 Irrigation system and management

The water used to supply the irrigation system came from a groundwater well belonging to the Agrometeorological Station of the Federal University of Ceará. The results of the water quality analysis for irrigation purposes are shown in Table 4. The results show that the water in question has high salinity (C3) and low sodium content (S1).

Table 4 - Water quality analysis results for irrigation purposes. Agrometeorological station, Fortaleza, Ceará, 2014.

Cations (mmolc L^{-1})	Anions (mmolc L^{-1})

Ca^{2+}	Mg^{2+}	In^{+}	K^{+}	Σ	Cl^{-}	SO?-	$HCO3^{-}$	$CO3^{2-}$	s
1,3	3,1	4,5	0,4	9,3	5,2	-	4,3	-	9,5
EC ($dS\ m^{-1}$)	RAS		pH		Dissolved solids ($mg\ L^{-1}$)			Classification	
0,95	3,05		8,0		950			C3 S1	

The drip irrigation system was adopted, using emitters with a variable flow rate of up to 10 L h^{-1}, which had previously been evaluated in the field, under normal operating conditions, according to the methodology described by Keller and Karmelli (1975). On this occasion, the irrigation system was tested using a nominal flow rate of 2 L h^{-1}, with a distribution uniformity coefficient (CUD) of approximately 90%.

The irrigation management method adopted was climate control. The methodology proposed by Bernardo, Soares and Mantovani (2006) was used to determine the potential crop evapotranspiration (ETpc), where the reference evapotranspiration (ETo) was first estimated.

To determine the daily estimate of reference evapotranspiration - Eto, a Class A evaporimeter tank was used, located close to the experimental area, in which the readings were taken daily at 9am.

Reference evapotranspiration (ETo) was estimated using equation 1 (BERNARDO; SOARES; MANTOVANI, 2006):

$$ETo = ECA \, . \, Kp \qquad (1)$$

in which,

ETo - Reference evapotranspiration estimated using the Class A Tank (mm);

ECA - Evaporation measured using the Class "A" Tank (mm);

Kp - Tank adjustment coefficient (dimensionless).

The equation proposed by SNYDER (1992), shown below, was used to obtain (kp):

$$Kp = 0{,}482 + 0{,}024\ln(F) - 0{,}000376\,U + 0{,}0045\,UR \qquad (2)$$

in which:

Kp - Tank coefficient;
F - Distance from the centre of the pond to the edge of the border (pond surrounded by grass at 10 m);
U - Average wind speed at 2 m height (km day$^{-1)}$;
RH - Average relative humidity (%).

The climate data needed to calculate (Kp) was obtained from the arithmetic mean of the last five years for the months in which the experiment was conducted, and was collected from an automated meteorological station located close to the experimental area.

Potential crop evapotranspiration (ETpc) was estimated using equation 3, by multiplying the reference evapotranspiration and the crop coefficient (Kc), according to Bernardo, Soares and Mantovani (2006).

$$ETpc = ETo.Kc \qquad (3)$$

in which,

ETpc - Potential crop evapotranspiration (mm);
ETo - Reference evapotranspiration estimated using the Class A Tank (mm);
Kc - Crop coefficients (0.70 up to 12 DAP; 0.81 from 13 to 33 DAP; 1.2 from 34 to 54 DAP and 0.77 from 55 DAP to the end of the cycle, as

proposed by Sousa, Bezerra and Teófilo (2005)).

The irrigation time used was calculated from equation 4:

$$T_I = \frac{ETpc \, . \, A_v}{CUD. \; q_g} \tag{4}$$

in which,

TI - Irrigation time (h);

ETpc - Potential crop evapotranspiration (mm);

Av - Vessel area (0, 0706 $m^{2)}$;

CUD - Water distribution uniformity coefficient of the irrigation system (Value obtained in evaluation, CUD = 0.90);

qg - Dripper flow rate, L h^{-1} (qg = 2 L $h^{-1)}$;

After planting the cowpea seeds, the experiment was irrigated daily with a rate equivalent to 100 per cent of the potential evapotranspiration of the crop - ETpc, in order to ensure the initial establishment of the crop, until the treatments began to differentiate, which took place 10 days after planting, by varying the salinity levels of the irrigation water.

After the irrigation water salinity treatments were imposed, the irrigation shift was changed to two consecutive days. The experiment was irrigated with a frequency equivalent to two consecutive days.

3.8 Salt concentrations in irrigation water

The saline water used for irrigation came from a groundwater well and was stored in reservoirs with a capacity of 310 litres. The amount of

NaCl, $CaCl_2.2H_2O$ and $MgCl_2.6H_2O$ salts used to prepare the irrigation water was determined in order to obtain the desired ECa in a 7:2:1 ratio, following the relationship between ECa and its concentration (mmolc L^{-1}= EC x 10), according to Rhoades; Kandiah; Mashali (2000).

3.9 Weed control and phytosanitary treatments

During the experiment, all the necessary cultural and phytosanitary treatments were carried out. Any weeds that occurred in the pots were eliminated manually in order to avoid competition with the bean plants for water and nutrients. Vertimec® was used to control aphids, at a rate of 1.5 mL per litre of water.

3.10 Variables analysed

At 45 days after planting, in order to measure the effects of the different treatments, the following variables were determined: plant height, stem diameter, number of leaves and leaf area.

3.10.1 Plant height

To determine the height of the plants, the distance between the neck of the plant and the youngest leaf was taken into account, measured using a graduated tape measure, with the value corresponding to the arithmetic mean of the six plants sampled, expressed in cm.

3.10.2 Stem diameter

Stem diameter was obtained by measuring the diameter at a distance of approximately 3 centimetres from the neck of the plant using a digital caliper, with the value corresponding to the arithmetic mean of the six plants sampled, expressed in mm.

3.10.3 Number of leaves

The number of leaves was obtained by counting them by hand from leaf no. 1 from the apex, with leaf no. 1 being the one that precedes the youngest leaf, and the value corresponding to the arithmetic mean of six plants sampled, expressed in units.

3.10.4 Leaf area

To determine leaf area, an area integrator (LI-3100, Li-Cor, Inc. Lincoln, NE, USA) belonging to the Laboratory of the Agrometeorological Station of the Federal University of Ceará was used.

leaves were detached from the stem and passed individually through the device, with the value corresponding to the arithmetic mean of six plants sampled, expressed in cm^2 $plant^{-1}$.

3.11 Statistical analyses

The data on the variables evaluated were analysed using the F test at 1 and 5% probability.

When a significant effect was found in the analysis of variance, the data obtained from the different treatments was subjected to a regression study (salinity levels) and a test of means (cultivars). The regression analysis sought to choose the mathematical model that best fitted the data from the experiment, taking into account the coefficients of determination (R2), the value of the F test from the analysis of variance and the aspects related to the biological behaviour of the processes, using, , the ASSISTAT (2015) statistical analysis software, aided by Microsoft Office Excel 2007.

CHAPTER 4

RESULTS AND DISCUSSION

The summary of the analyses of variance for the morphological data of plant height (h), stem diameter *(0c),* number of leaves (N_f) and leaf area (AF) of the caupi bean can be seen in Table 5. There was a significant influence at the 1% probability level by the F test in the salinity level treatments on the morphological variables plant height (h), stem diameter (0c), number of leaves (Nf) and leaf area (Af). The cultivars were significant for plant height at the 1% level and for the stem diameter and leaf area variables at the 5% level. There was an interaction between the treatments for the stem diameter variable at the 5% level.

Table 5 - Summary of analysis of variance for plant height (h), stem diameter (0c), number of leaves (Nf) and leaf area (Af) of cowpea subjected to salinity levels, Fortaleza, Ceará 2014

Source of variation	GL	Mean squares			
		h (cm)	Øc (mm)	Nf	Af ($cm^2 plant^{-1}$)
Blocks	2	1.23^{ns}	0.14^{ns}	7.90^{ns}	70898,44**
Salinity levels (N)	4	114,88**	11,21**	90, 41**	1710424,25**
Residue (N)	8	0,48	0,11	10, 44	7305,38
Parcels	14	-	-	-	-
Cultivars (C)	1	38,53**	0,99*	4.80^{ns}	363636,07*
Int. N x C	4	1.28^{ns}	0,76*	5.71^{ns}	45327.83^{ns}
Residue (C)	10	2,63	0,15	2,93	41180,11
Total	29				

GL - Degrees of freedom; CV - Coefficients of variation; ^{ns} - Not significant, - Significant at 1% probability by the F test and * - Significant at 5% probability by the F test.

Analysing the effect of salinity levels on the variable plant height (h) showed that the mathematical regression model that best fitted the data for this variable was linear, with a value of 0.96 for the coefficient of determination ($R^{2)}$ (Figure 8).

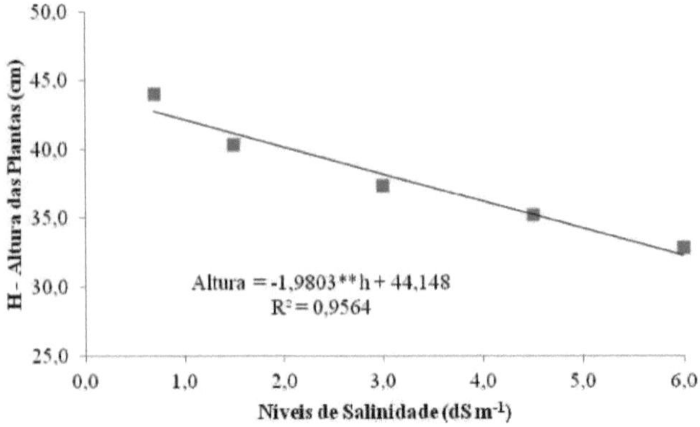

Figura 8 - Plant height (cm) as a function of salinity levels, Fortaleza, Ceará, 2014

This negative influence of salinity levels on plant height may be related to the action of the osmotic component that makes it difficult for the plant to absorb water, as well as the toxicity caused by an excess of certain ions, as stated by Coelho et al. (2013).

Working with cowpeas irrigated with saline water, Silva et al. (2011) observed that increasing levels of salt in the irrigation water affected plant height growth. Lima et al. (2007), evaluated the effect of salinity on the development of cowpea plants and found a reduction of around 42.86% in height when irrigated with the highest level of saline water, while this study found a reduction of 25.39%. However, Veras et al. (2011), evaluating growth and production parameters of jatropha irrigated with saline water, found that plant height was not affected by water salinity, demonstrating the tolerance of jatropha to salinity; however, the same characteristic cannot be attributed to cowpea.

Figure 9 shows the behaviour of the cultivars in terms of plant height. The results show that the Epace 10 cultivar was 5.88 per cent taller than BRS Itaim.

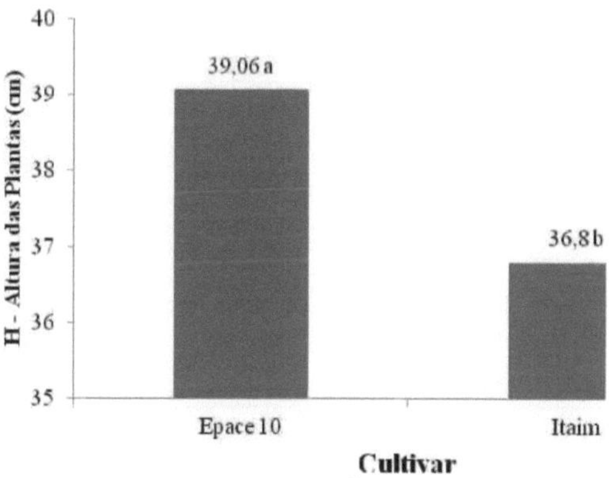

Figura 9 - Average plant height values of the cultivars analysed, Fortaleza, Ceará, 2014

As for the effect of salinity levels on stem diameter (0c), regression analysis revealed that the mathematical model that best represented the data for this variable was the linear type, with a coefficient of determination (R^{2}) of 0.99 for the BRS Itaim cultivar and (R^{2}) of 0.87 for the Epace 10 cultivar (Figure 10).

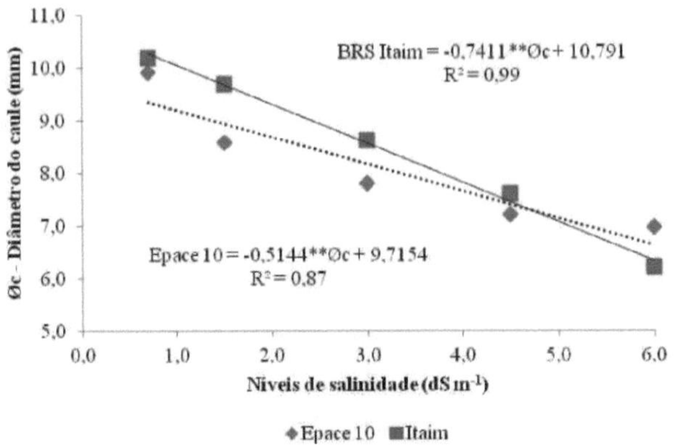

Figura 10 - Stem diameter of the Epace 10 and BRS Itaim varieties as a function of salinity levels. Fortaleza, Ceará, 2014

The stem diameter of the cowpea plants was negatively affected by the increase in saline water, with greater intensity in the treatments with the highest salinity levels. However, it was the characteristic least affected by the saline conditions studied, with a reduction of 39.05% and 29.63% in relation to the treatments with the lowest salinity (0.7 dS m^{-1}) and the highest salinity (6.0 dS m^{-1}) in the BRS Itaim cultivar and the Epace 10 cultivar, respectively. The literature has shown that this species tolerates saline water with an electrical conductivity of up to 3.3 dS m^{-1} and is classified as moderately tolerant (AYERS; WESTCOT, 1999).

The reduction in caupi stem diameter when subjected to saline stress has been observed in several scientific studies. Lima et al. (2007) observed a 34.69% reduction in relation to the treatment with the lowest salinity in their study with the quarentinha cultivar. Nascimento et al. (2013) observed a 13.46 per cent reduction when compared to the highest salinity level (5.1 dS $m^{-1)}$. In contrast to these authors, Dantas (2014) observed that the stem diameter of peppers increased as salinity levels increased. This response can be attributed to the osmotic adaptation capacity that some crops have, allowing them to absorb greater amounts of water, even in a saline environment (AYERS; WESTCOT, 1991).

When analysing the effects of salinity levels on the variable number of leaves, it was found that the mathematical regression model that best represented the plant's response was the linear one, with a value of 0.92 for the coefficient of determination (R^2) (Figure 11).

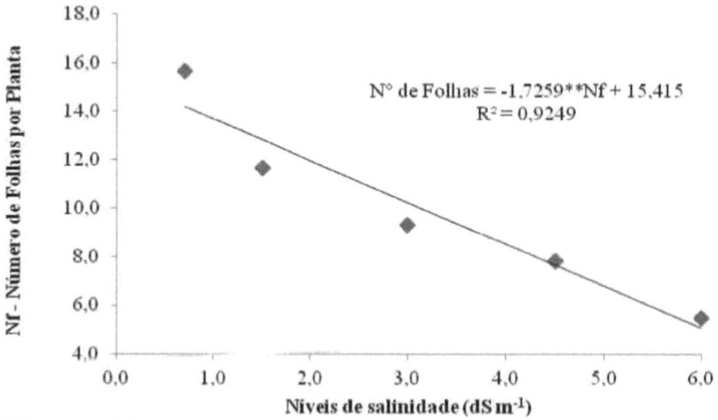

Figure 11 - Number of leaves of the varieties Epace 10 and BRS Itaim as a function of salinity levels, Fortaleza, Ceará, 2014

Figure 11 shows a decrease in the number of leaves as the salinity levels in the irrigation water increased, with a reduction of 64.74 % at the highest salinity level (6.0 dS m^{-1}).

This reduction in the number of leaves for the higher level of salinity must be related to the plant's possible adaptation mechanisms to salt stress, consisting of a decrease in the production of leaf area, the closure of stomata, accelerated senescence and leaf abscission, limiting not only the size of individual leaves, but also the number of leaves by reducing the number and growth rate of branches (TAIZ; ZEIGER, 2009).

Due to the accumulation of abscisic acid, induced by stress conditions, both water and salt, there is inhibition of cell division and consequent reduction in leaf area, while leaf abscission occurs due to ethylene synthesis, which is high in plants exposed to these stresses (TAIZ; ZEIGER, 2009).

Work carried out with other crops has also demonstrated the effect of salinity on the number of leaves. Oliveira et al. (2006) and Oliveira et al. (2007) observed a reduction in the number of leaves in castor bean and corn-popcorn crops, respectively, as the saline concentration of the irrigation water increased.

For the effect of salinity levels on the leaf area variable (Af), regression analysis showed that the mathematical model that best represented the data for this variable was linear, with a coefficient of determination (R^2) of 0.89 (Figure 12).

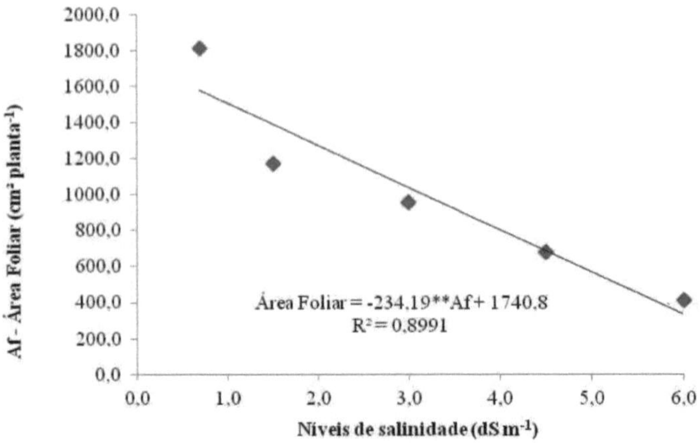

Figura 12 - Leaf area of the varieties Epace 10 and BRS Itaim as a function

of salinity levels, Fortaleza, Ceará, 2014

Figure 12 shows a decrease in leaf area as the salinity levels in the irrigation water increased, with a 77.34 % reduction at the highest salinity level (6.0 dS).

It is likely that the inhibition of leaf area was mostly caused by the toxic effects of the salts absorbed by the plants, by the low osmotic adjustment capacity of the crop or by the reduction in total water potential caused by the increase in salt concentration, generating a physiological disturbance in the plant and compromising the opening of the stomata (LARCHER, 2006).

Similar to this result, Araújo et al. (2010) observed in their experiment with bean plants that the leaf area of the four varieties studied decreased as salinity increased. In studies with tomatoes, Sanchez Blanco et al. (1991) found a greater reduction in the leaf area produced at high salinity levels.

The average values for the leaf area variable can be seen at in figure 13. The results show that the Epace 10 cultivar was 19.74 per cent superior to BRS Itaim.

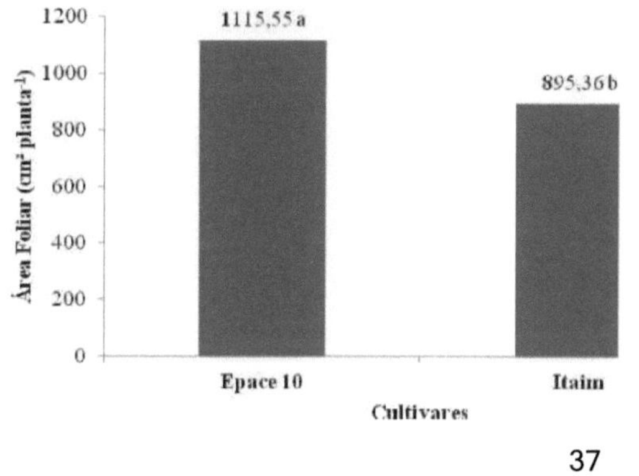

Figura 13 - Average values for the leaf area variable of the cultivars analysed. Fortaleza, Ceará, 2014

CHAPTER 5

CONCLUSIONS

1. The morphological characteristics of the cowpea (plant height, stem diameter, number of leaves and leaf area) were reduced as the salinity of the irrigation water increased.
2. The Epace 10 cultivar showed the highest values for height and leaf area.

REFERENCES

AQUINO, A. B. AQUINO, B. F.; HERNANDEZ, F. F. F.; HOLANDA, F. J. M.; FREIRE, J. M.; CRISOSTOMO, L. A.; COSTA, R. I.; UCHÔA, S. C. P.; FERNANDES, V. L. B. **Fertilisation and liming recommendations for the state of Ceará**. Fortaleza: UFC, 1993. 247p.

ARAÚJO, **J. P. P. de; SANTOS, A. A. dos; CARDOSO, M. J.; WATT, E. E.** Note on the occurrence of a branched inflorescence in caupi Vigna unguiculata (L.) Walp. Subsp, unguiculata in Brazil. **Revista Ciência Agronômica**, v.12, n.1/2, p.187-193, 1981.

ARAÚJO, C.A.S.; RUIZ, H.A.; CAMBRAIA, J.; NEVES, J.C.L.; FREIRE, M.B.G.S.; FREIRE, F. J. Varietal selection of Phaseolus vulgaris for salt stress tolerance based on growth variables. **Rev. Ceres**, v.57, n. 1, p. 2-4, 2010.

ASSISTAT. Version 7.5 beta By Francisco de A. S. e Silva. DEAG-CTRN- UFCG - Updated on 04/01/2015. Available at <HTTP://www.assistat.com> Accessed on 17/06/2015.

AYERS, R. S.; WESTCOT, D. W. **Water quality in agriculture**. Translation by GHEYI, H. R.; MEDEIROS, J. F.; DAMASCENO, F. A. V. Campina Grande: UFPB, 1991. 218p. (FAO. Estudos de Irrigação e

Drenagem, 29).

AYERS, R.S.; WESTCOT, D.W. **The quality of water for irrigation.** Translation by GUEYI, H.R.; MEDEIROS, J.F.; DAMASCENO, F.A.V. Campina Grande: UFPB, 1999. 153p. (FAO. Irrigation and Drainage, 29. Revised).

BERNARDO, S.; SOARES, A. A.; MANTOVANI, E. C. **Irrigation Manual.** 8. ed. Viçosa, MG: UFV, 2006. 625 p

BOHN, H.L.; McNEAL, B.L.; O'CONNOR, G.A. **Soil Chemistry.** New York. p.329.1979

CARDOSO, J. M.; MELO, F. B.; LIMA, M. G. Ecophysiology and planting management. In: FREIRE FILHO, F. R.; LIMA, J. A. A.; RIBEIRO, V. Q. **Feijão caupi: avanços tecnológicos.** Brasília: EMBRAPA, 2005. Chap. 5, p 213 - 225.

COELHO, J. B. M.; BARROS, M. F. C.; BEZERRA NETO, E.; CORREA, M. M. Water behaviour and growth of vigna beans grown in salinised soils. **Brazilian Journal of Agricultural and Environmental Engineering,** v.17, p.379-385, 2013.

COSTA, P. H. A.; SILVA, J. V.; BEZERRA, M. A.; ENEAS FILHO, J.; PRISCO, J. T.; GOMES FILHO, E. Growth and levels of organic and inorganic solutes in cultivars of Vigna unguiculata submitted to salinity. **Revista Brasileira de Botânica**, v. 26, n.3, p.289-297, 2003.

DADSON, R. B.; HASHEM, F. M.; JAVAID, I.; JOSHI, J.; ALLEN, A. L.; DEVINE, T. E. Effect of water stress on the yield of cowpea [Vigna unguiculata (L.) Walp] genotypes in the Delmarva region of the United States. **Journal Agronomy & Crop Science**, v.191, p. 210-217, 2005.

DANTAS, G. F. P. **Effect of different levels of salinity in irrigation water on pepper cultivation.** 2014. 30 f. Monograph (Graduation in Agricultural Sciences), State University of Paraíba, Centre for Agricultural Sciences, 2014.

DIAS, N. S.; BLANCO, F. F. Effect of salts on soil and plant. In: GHEYI, H. R.; DIAS, N. S.; LACERDA, C. F. **Management of salinity in agriculture: basic and applied studies.** Fortaleza: INCTsal, 2010. 472 p.

BRAZILIAN AGRICULTURAL RESEARCH COMPANY - EMBRAPA. **Brazilian soil classification system.** 3.ed. Brasília, 2013. 353p.

FREIRE FILHO, F. R.; RIBEIRO, V. Q.; BARRETO, P. D.; SANTOS, A. A. dos. **Cowpea cultivars for the Mid-North region of Brazil.** In: CARDOSO, M. J. (Org.) A cultura do feijão-caupi no Meio-Norte do Brasil. Teresina: Embrapa-CPAMN, 2000. p.67-68 (Embrapa-CPAMN, Technical Circular).

GHEYI, H.R.; MEDEIROS, J.F.; BATISTA, M.A.F. **Prevention, management and recovery of saline and sodic soils.** Specialisation Course in Irrigation and Drainage. Mossoró: ESAM, 1991. 56p. Handout.

GHEYI, H. R.; CORREIA, K. G.; FERNANDES, P. D. **Soil salinity and plant growth and development.** In: NOGUEIRA, R. J. C.; ARAÚJO, E. L.; WILLADINO, L. G.; CAVALCANTE, U. M. T. Estresses Ambientais: danos e benefícios em plantas. Recife: UFRPE: University Press, 2005, p. 95-105.

HOLANDA, J. P.; AMORIM, J.R.A. **Water quality for irrigation.** In: GHEYI, H.R.; QUEIROZ, J.E.; MEDEIROS, J.M. (eds) Manejo e Controle da

salinity in irrigated agriculture. Campina Grande: UFPB-SBEA, 1997. Chap.5, p.137-169.

KELLER, J.; KARMELI, D. **Trickle irrigation design.** S.1: Rain Bird Sprinkler Manufacturing Corporation, 1975. 133 p.

LACERDA, C. F. Effects of salinity on the development and mineral composition of string beans (Vigna unguiculata (L.) Walp.) and the use of Ca^{2+} as a means of minimising these effects. 1995. 87 f. Dissertation (Master's degree in soils and plant nutrition). Federal University of Ceará,

Fortaleza, 1995.

LARCHER, W. **Plant Ecophysiology.** Translation: Prado, C. H. B. A. São Carlos: Ed. Rima, 2006, p. 531.

LIMA, C. J. G. S.; OLIVEIRA, F. A.; MEDEIROS, J. F.; OLIVEIRA, M. K. T.; ALMEIDA JÚNIOR, A. B. Response of cowpea to irrigation water salinity. **Revista Verde,** v. 2, n. 2, p. 79-86, 2007.

MAAS, E. V.; HOFFMAN, G. J. Crop salt tolerance: current assessment. **Journal of the Irrigation and Drainage Division**, A.S.C.E., v. 103, n.2, p.115- 134, 1977.

MEDEIROS, J. F.; GHEYI, H. R. **The quality of irrigation water.** Campina Grande: UFPB, 1994. 60p.

MORAIS, F. A. de; GURGEL, M. T.; OLIVEIRA, F. H. T. de; MOTA, A. F. Influence of irrigation with saline water on sunflower cultivation. **Revista Ciência Agronômica,** v. 42, n. 2, p. 327-336, 2011.

MOUSINHO, F. E. P. **Economic viability of cowpea irrigation in the state of Piauí.** 2005. 125p. Thesis (Doctorate). Piracicaba, SP. University of São Paulo.

NASCIMENTO, R.; ANDRADE, J. R.; JUNIOR, S. O. M.; BARBOSA, J. W. S.; ALENCAR, A. E. V.; SOUSA, J. S. Growth of cowpea genotypes irrigated with saline water. **Revista Agropecuária Científica no Semiárido**, v. 9, n. 4, p. 41 - 45, 2013.

OLIVEIRA, M. K. T.; OLIVEIRA, F. A.; MEDEIROS, J. F.; LIMA, C. J. G. S.; GUIMARÃES, I. P. Effect of different bovine manure contents and salinity levels on the initial growth of papaya (Ricinus communis). **Magazine**

Green of Agroecology and Sustainable Development, Mossoró, v.1. n.1, p.68-74, 2006.

OLIVEIRA, F. A.; MEDEIROS, J. F.; OLIVEIRA, M. K. T.; LIMA, C. J.

G. S.; GALVÃO, D. C. Initial development of 'Jade' popcorn irrigated with water of different salinity levels. **Revista Verde de Agroecologia e Desenvolvimento Sustentável**, Mossoró, v.2, n.1, p.45-52, 2007.

REED, D. W. Combating poor water quality with water purification systems. In : Reed, D. W. (ed) **Water , media and nutrition for greenhouse crops**. Illionois: USA, Ball Publishing, 1996, p.51-67.

RHOADES, J. P.; KANDIAH, A.; MASHALI, A. M. **The use saline waters for crop production.** Trad.: GHEYI, H. R.; SOUSA, J. R.; QUEIROZ, J. E. Campina Grande: UFPB, 2000. 117p.

SANCHEZ BLANCO M. J. ; BOLARIN M. C, ALARCON JJ & TORRECILLAS A. Salinity effects on water relations in Lycopersicon esculentum and its wild salt-tolerant relative species L. pennellii. **Physiologia Plantarum**, v. 83:p. 269-274, 1991.

SILVA, J. V. Effects of CaCl2 on growth and accumulation of osmoregulators in string bean plants cv. Pitiúba stressed with NaCl. 1998. 103p. Dissertation (Master's in Plant Science) - Federal University of Ceará, Fortaleza, 1998.

SILVA, J. V.; LACERDA, C. F.; COSTA, P. H.; ENÉAS FILHO, J.; GOMES FILHO, E.; PRISCO, J. T. Physiological responses of NaCl stressed cowpea plants grown in nutrient solution supplemented with CaCl2. **Brazilian Journal of Plant Physiology**, v. 15, p. 99-105, 2003.

SILVA, F. L. B.; LACERDA, C. F.; SOUSA, G. G.; NEVES, A. L. R.; SILVA, G. L.; SOUSA, C. H. C. Interaction between salinity and bovine biofertiliser in cowpea cultivation. **Revista Brasileira de Engenharia Agrícola e Ambiental**, v. 15, n. 4, p.383-389, 2011.

SINGH, B. B.j EHLERS, J. D.; SHARMA, B.; FREIREFILHO, F. R. **Recent progress in cowpea breeding.** In: FATOKUN, C. A.; TARAWALI, S. A.; SINGH, B. B.; KORMAWA, P. M.; TAMO, M. (Eds) Challenges and Opportunities for Enhancing Sustainable Cowpea Production. International Institute of Tropical Agriculture, Ibadan, Nigeria, p.22-40, 2002.

SMARTT, J. **Evolution of genetic resources.** In: SMARTT, J. (Ed.), Grain Legumes. Cambridge University Press, Cambridge, p.140-175, 1990.

SNYDER, R. L. Equation for evaporation pan to evapotranspiration conversion. **Journal of Irrigation and Drainage Engineering**, v.118, n.6, p.977-980, 1992.

TAIZ, L.; ZEIGER, E. **Plant Physiology.** 4.ed. Porto Alegre: E. Atmed, 2009. 819 p.

TÉOFILO, E. M.; MAMEDE, F. B.; SOMBRA, N. S. Natural hybridisation in cowpea. **Ciência e Agrotecnologia**, Lavras, v.23, n.4, p.1011-1012, 1999.

TRAVASSOS, K. D.; GHEYI, H. R.; SOARES, F. A. L.; BARROS, H. M. M.; DIAS, N. S.; UYEDA, C. A.; SILVA, F. V. Growth and development of sunflower varieties irrigated with saline water. **Revista Irriga**, special issue, p.324 - 339, 2012.

TYERMAN, S. D. & SKERRETT, L. M. Root ion channels and salinity. **Scientia Horticulturae**, v. 78, p. 175-235, 1999.

VERAS, R. P.; LAIME, E. M. O.; FERNANDES, P. D.; SOARES, F. A. L.; FREIRE, E. A. Plant height, stem diameter and yield of jatropha irrigated under different salinity levels. **Revista Brasileira de Engenharia Agrícola**, v.15, n. 6, p 582-587, 2011.

YAHYA, A. Salinity effects on growth and on uptake and distribution of sodium and some essential mineral nutrients in sesame. **Journal of Plant Nutrition**, v.21, n.7, p. 1439-1451, 1998.

YEO, A. R. Predicting the interaction between the effects of salinity and climate change on crop plants. **Science Horticulturae**, v.78, p. 159-174, 1999.

MIX
Papier aus verantwortungsvollen Quellen
Paper from responsible sources
FSC® C105338

Printed by Books on Demand GmbH, Norderstedt / Germany